MY VERY OWN® NAME

This book was created especially for
Maximilian Smith

2011

Written by Maia Haag • Illustrated by Mark Mille • Designed by Haag Design, Inc.

PERSONALIZED CHILDREN'S BOOKS

On May 29, 2010, the news spread far and wide.
"Did you hear?" the creatures said, "A baby boy arrived."

Animals came to see the sight. They gathered all around.
"We must create a name for him that has a perfect sound."

"To spell the name we need to
choose the letters that we like,"
the wise owl said while looking
at the baby with delight.

The moose snorted and started to grunt.
"The letter **M** is the one that I want!"

Ma

The antelope said,
"It would be a shame
if **A** were missing
from the name."

Max

"The letter **X** would be my wish!"
said the nearly transparent x-ray fish.

Maxi

While jumping more than eight feet high,
the impala suggested the letter **I**.

Maxim

Scratching her head
and laughing with glee,
the monkey tossed down
an **M** from her tree.

Maximi

The Irish Setter
wagged his long tail.
"I've found an **I**,
and I'm on its trail!"

Maximil

"BAAA!" said the lamb from her mommy's lap.
"Can we use an **L** before I nap?"

Maximili

"Please, please, can we use an **I**?"
the inchworm squeaked with a tiny sigh.

Maximilia

The anteater said while lifting his snout,
"**A** is the very best letter, no doubt!"

Maximilian

The nightingale sang with a tear in her eye,
"If **N** is not used, I do not want to fly!"

The wise old owl who watched from the trees
flew down to say that he was pleased.

"Maximilian is a very good start;
but, my friends, before we part,
we need to spell the last name next.
Let's add it now before we rest."

S

The squirrel yelled from the lofty treetop,
"I'm bringing an **S** that I don't want to drop!"

Sm

"I know that this letter might look the same,
but can we use it again to spell the name?"

Smi

The iguana uttered while creeping along,
"I think this is where an **I** should belong!"

Smit

"If we use a **T**,
    it will be such a thrill!"
the toucan squawked
        with his colorful bill.

Smith

While balancing an **H** on her spotted calf,
the silly hyena did nothing but laugh.

"Maximilian Smith has a wonderful sound!
A more perfect name we could not have found.

Now that our work for the day is complete,
Let's sing and dance and stomp our feet!"

"The story is over, but do not be sad,
for now it is time for more fun to be had.

I have listed the animals
so you can look
to see if you find them
throughout the book."

Animal Encyclopedia

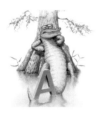

Alligator
I am a powerful reptile that lives by the water. Watch out for my sharp teeth! I have scales all over my body.

Ant
I live in an underground nest with many other ants. I carry food back to the nest, and sometimes I get tired!

Anteater
Can you guess what my favorite food is? Ants! I stick out my long tongue to pick up ants and eat them.

Antelope
I like to swallow grass and then bring it back up into my mouth to chew it again. I know this is bad manners!

Bear
Grrr! I am stronger and taller than a person. At the end of the summer, I take a nap, and I sleep until spring.

Beaver
I have sharp teeth that I use to cut wood. I make my home, called a "dam," out of branches in a lake.

Caterpillar
I have lots of feet. I love to eat leaves. I will wrap myself in a cocoon and then turn into a beautiful butterfly!

Cheetah
I can run faster than a car on a highway. My tail helps me keep my balance when I run so fast.

Deer
I am a shy animal that eats grass, weeds, twigs, shrubs and nuts. At night, I make a soft bed in the grass.

Donkey
Hee-haw! My voice is loud, and my ears are big. They help me hear other donkeys a long way away.

Eagle
Did you know my wings can be eight feet wide, which is as wide as a car? My nest is so big a person could lie in it!

Elephant
I am the largest animal on land. I can weigh up to 13,000 pounds! I fan my big ears to cool my body.

Elk
I am a male elk with strong antlers. I grow a mane of long hair on my neck to keep me warm in the winter.

Emu
I'm a mommy emu. When I lay eggs, the daddy emu sits on them to keep them warm. He's careful not to break them!

Flea
I am so small you can hardly see me. I have strong legs so I can jump high. I live in the fur of animals and birds.

Inchworm
It takes me a long time to move one inch! I am small, and I eat lots of leaves. I will turn into a brown moth.

Frog
Ribbit! Ribbit! I used to be a tadpole with a long tail and fins. Then I turned into a frog with long legs.

Irish Setter
Ruff! I love to play, and I'm a great hunter. Sometimes I hunt in the field, and sometimes I hunt for treasures in the closet.

Giraffe
My legs and neck are so long that I am as tall as a tree! I love to stretch my neck to eat the leaves.

Jackal
I look a little bit like a fox. When I am hungry, I eat the food left by other animals such as lions and tigers.

Gorilla
Did you know I can weigh as much as 595 pounds? I walk on the knuckles of my hands and feet.

Jaguar
My body is strong. I like to climb trees and lie on branches. I am also an excellent swimmer.

Hippopotamus
When I walk on land, you can see my big mouth and belly. I can walk along the bottom of rivers and lakes.

Kangaroo
My baby kangaroo lives in my pouch for the first eight months of her life. I move by hopping. I live in Australia.

Hyena
I am a very noisy animal! My howling screams sound like laughter. I often run with a group or "pack" of hyenas.

Koala
When I was a baby, I lived in my mom's pouch at first, and then I rode on my mom's back. I live in Australia, too.

Iguana
I live in trees by water, and I can swim. I have green bands across my shoulders and a tail that becomes darker with age.

Ladybug
My red color and black spots warn birds that I would not make a tasty meal. I think my spots are so pretty!

Impala
I can leap as high as 10 feet up in the air and as far as 33 feet! I live in herds of over 200 impalas.

Lamb
I am a baby sheep, and this is my mom. I was born in the spring. When my hair is cut, the wool is used for sweaters.

Lion
Roar! I am strong, and I am the king of the jungle. I have a thick mane of hair. My baby lions are called cubs.

Monkey
I use my long tail to swing from branch to branch in trees. I might look silly, but I am very smart!

Moose
Did you know I have a flap of skin called a "bell" that hangs from my throat? I am a great swimmer.

Nag
"Nag" is another word for "old horse." Do I look old? I like to eat grass, and I like to walk around in the field.

Newfoundland
I will grow up to be a large, strong dog. I'm such a good worker dog that I can pull nets for fishermen.

Nightingale
Fa, la, la, la, la! I sing from early in the morning until the end of the day. Sometimes I sing until midnight!

Orangutan
My arms are so long they reach my ankles when I am standing up. I sleep on a flat nest of sticks in a tree.

Ostrich
I live on the ground even though I am a bird. One of my eggs is the size of 40 hen eggs put together!

Otter
I love to play and swim. Did you know I can close my ears and nose when I am in the water?

Peacock
Aren't I handsome? I spread my pretty tail to show it to peahens. I eat grain, fruit, berries and insects.

Porcupine
Watch out for my sharp spines! You might see me at night. I walk on the ground, and I can climb trees.

Quail
I am hard to find because I live under small bushes. While I can fly, I usually run through plants instead.

Rabbit
My fur is very soft. I hop using my long back legs. Sometimes I am naughty and eat plants in the garden.

Rhinoceros
When it's hot, I love to lie in the mud. All my friends lie in the mud, too. We don't mind getting dirty.

Rooster
Cock-a-doodle-do! I am a male chicken. I crow very loudly, very early in the morning. I wake people up!

Skunk
I can make a really bad smell so my enemies will go away. My white stripe tells animals to stay back.

Snail
My eyes are on long stalks. I have a slimy body, inside a hard shell. It takes me a long time to move anywhere.

Squirrel
My favorite way to play is to chase my friends through the treetops. Have you ever seen me jumping up high?

Tiger
I am a powerful, very large cat. There are stripes all over my body. I am a fast runner and a good climber.

Toucan
I live in the rainforest in South America. I like to play with other toucans by tossing berries into their beaks.

Turtle
I have a shell made of scales of bone. I can pull my head into my shell. Please be careful not to step on me!

Umbrella Bird
My "umbrella" feathers help female birds notice me. I live in the tops of trees. I spring noisily through the branches.

Unicorn
I am an imaginary animal that lives in fairy tales. I look like a horse, but I am white, and I have a long horn.

Viper
Hiss! Hiss! I am a snake that can be as long as six feet. I have a beautiful pattern on my back.

Vole
I am a mouse that lives in the country. I make tunnels in the long grass and dig in the ground. Aren't I cute?

Warthog
I wish I had a prettier name than warthog! I like to wade in water and rest in holes among rocks. I live in Africa.

Wolf
Howooo! I make a howling noise that sounds like a loud cry. I look a little bit like a dog, but I live in the wild.

X-Ray Fish
Parts of my body are clear, so you can see through them. I can lay 400 eggs at a time. I live in South America.

Yak
I live on mountains in cold places. My long hair almost reaches the ground. I never comb my messy hair!

Yellow Jacket
Buzz! Buzz! I am a wasp. I look as if I'm wearing a yellow jacket. Watch out for my sting!

Zebra
I am covered in stripes. Did you know that I can recognize other zebras in my family by their look, sound and smell?

PERSONALIZED CHILDREN'S BOOKS

www.iseeme.com
1.877.744.3210 (toll-free)

See all the titles in our *My Very Own®*
personalized book series:

- *My Very Own Name*
- *My Very Own Fairy Tale*
- *My Very Own Pirate Tale*
- *My Very Own ABC's*

I See Me! Inc.
14505 27th Ave. North
Plymouth, MN 55447